青岛大剧院
The Qingdao
Grand Theater
in China

主编

（德）曼哈德·冯·格康
Meinhard von Gerkan
（德）斯特凡·胥茨
Stephan Schütz

辽宁科学技术出版社
·沈阳·

图书在版编目（CIP）数据

青岛大剧院 /（德）格康，（德）胥茨主编. -- 沈阳:
辽宁科学技术出版社，2014.5
　ISBN 978-7-5381-8565-2

Ⅰ. ①青… Ⅱ. ①格… ②胥… Ⅲ. ①剧院－建筑设
计－青岛市 Ⅳ. ①TU242.5

中国版本图书馆CIP数据核字(2014)第065426号

主　　编：曼哈德·冯·格康（gmp）、斯特凡·胥茨（gmp）
总　策　划：迈克尔·库恩（gmp）、柳青（UED）
策　　划：克劳迪娅·苔斯勒（gmp）、方小诗（gmp）、郑珊珊（gmp）
版式设计：汤姆·魏伯伦茨、亨德里克·西什莱、欧恩平面设计公司（德国汉堡）
封面设计：欧恩平面设计公司（德国汉堡）
公司地址：德国汉堡易北大道139,22763
网　　址：www.gmp-architekten.de　www.uedmagazine.net

出版发行：辽宁科学技术出版社
　　　　　（地址：沈阳市和平区十一纬路29号 邮编：110003）
印　刷　者：北京雅昌彩色印刷有限公司
幅面尺寸：190mm×292mm
印　　张：5
字　　数：12千字
出版时间：2014年6月第1版
印刷时间：2014年6月第1次印刷
责任编辑：姜思琪　包伸明
责任校对：王玉宝

书　　号：ISBN 978-7-5381-8565-2
定　　价：149.00元

青岛大剧院

The Qingdao Grand
Theater in China

gmp·德国冯·格康，玛格及合伙人建筑师事务所

地理位置

历史悠久的海滨及港口城市青岛位于山东半岛东南部，其三面为黄海包围。因其地理位置，在19世纪末发展成为重要的商贸城市。德国殖民时代对青岛的影响包括举世闻名的青岛啤酒，其严格按照德国啤酒《纯净法》酿制。而直到今天，这座拥有800万人口的大都会中，具有德国风格的建筑仍随处可见。青岛因其历史原因与国际间交流密切。

青岛地理位置得天独厚，山海形胜，腹地广阔，位于崂山海拔1133m高巨峰和黄海绵长的海岸线之间。青岛作为深受喜爱的海滨城市在德国享有"黄海边的那不勒斯"之誉。2008年奥运会举办期间，浮山湾成为帆船赛事的举办场地。青岛港作为中国第三大深水港赋予城市蓬勃的经济活力。

Location

One of China's most popular seaside resorts and the nation's third largest harbor city overlooks the Yellow Sea on three sides. Due to its outstanding geographical location, Qingdao has been a busy trading post since the end of 19th century. International influences can also be detected: for example, the world famous Tsingtao-beer is brewed there according to the German beer brewing tradition. Furthermore, the city center is adorned by quite a few buildings that emulate indigenous German architecture. Qingdao has grown into a metropolis of 8.5 million, which is still proud of its historical, cultural, and architectural heritage.

Geographically, Qingdao is situated between the picturesque, 1,133-meter-high Laoshan mountain range and the wide beaches of the Yellow Sea. It is therefore no surprise that in Germany, China's seaside resort is praised as the "Naples on the Yellow Sea". In 2008, the Olympic sailing competitions were held in the bay of Fushan. Economically, it is the deep water harbor that bestows an even greater significance and power to Qingdao than its cultural heritage.

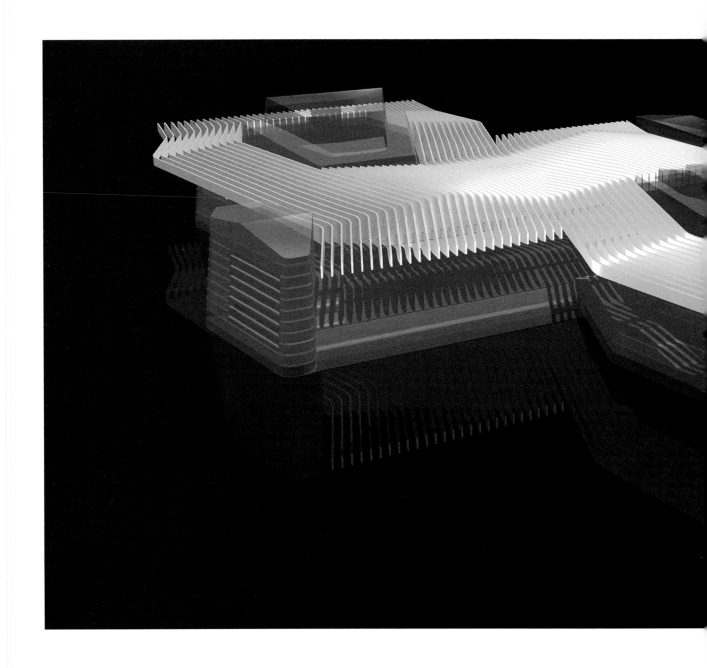

"云雾屋面"数字模型。屋面跨度达60m，个别屋面百叶构件的结构高度达6m，其
向两侧边缘处高度逐渐减小至2m。
Digital model of the "cloud roof". The span of the roof extends up
to 60 meters; its vertical panels are up to 6 meters high, reducing to
2 meters at the edge of the roof.

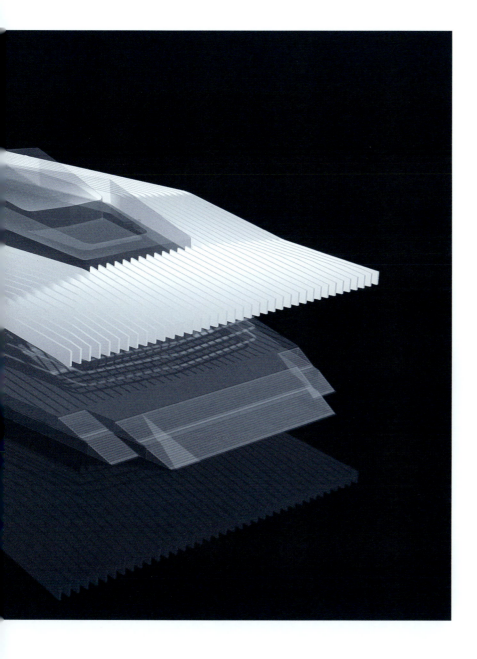

西侧建筑体内的歌剧厅设有1 600个坐席，主幕和侧幕均采用了最现代的科技技术。大型室外阶梯面向南侧的大海开放，而位于"山峰"内部的歌剧厅在晚上将由节庆般的红色灯光照亮。歌剧厅采取了欧洲歌剧院的经典格局，位于多层之上的台座以马蹄铁的形式环绕舞台布置。大厅内壁以彩色混凝土镶板包裹造型，符合对声学的要求，保证了剧院可以举行高标准的国际演出。

东侧建筑体内设有音乐厅和多功能演出厅。其"脊背相抵"的布局方式实现了共享上货区和更衣室的功能要求。音乐厅内设有1 200个坐席，同样严格符合高标准的声学要求。多边形的大厅吊顶和屋面均由流线型暖色榆木镶板覆盖，呈现简洁的白色。波浪状的造型除符合声学要求外，还传达了灵动起伏的音乐海洋的意象。

所有演出大厅的外部幕墙、前厅部分以及云雾屋面上均布有灯光照明带，弱化了建筑的内外边界：令人们在建筑内部仍如同徜徉于山海之间。

The Opera House, with its 1,600 seats and state-of-the-art technology serving its main and secondary stages, is situated in the western part of the building. The large open-air flight of steps opens up to the sea to the south, and within the "mountain", the festive red of the auditorium shimmers at night. The auditorium replicates the design of classic European opera houses, with numerous balconies surrounding the stalls in the shape of a horseshoe. The sculptural cladding with colored concrete panels fulfills complex acoustic requirements.

The eastern part of the building accommodates the concert hall and the multifunctional hall. This back-to-back positioning makes it possible to share the central cloakroom and delivery facilities. The concert hall with seating for 1,200 also complies with top acoustic requirements. The diamond-shaped hall comprises curved stalls made of warm elm wood and wall and ceiling areas in pure white. Again, the sculptural, wave-shaped design of the components is important for the acoustics.

The façades of the performance areas also enclose the foyers, and the "cloud roof" is extended in lamella-shaped cantilevers so that the boundaries between inside and outside are blurred – the walk through the landscape of the Laoshan mountain range continues on the inside.

2005年前期设计阶段效果图
Visualization from the 2005 outline design phase

↗ 功能层平面 +-0.00m
Layout drawing of functional floor level +- 0.00

1	歌剧厅舞台	Opera stage
2	排练厅	Rehearsal rooms
3	乐器博物馆	Musical Instruments Museum
4	餐厅	Restaurant
5	宴会厅	Banquet hall
6	会议厅	Conference halls
7	酒店大堂	Hotel lobby

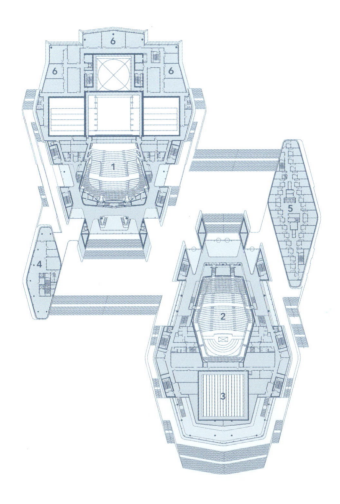

↗ 演出大厅层平面 + 9.00 m
Layout drawing at auditorium level + 9.00 m

1	歌剧厅	Opera hall
2	音乐厅	Concert hall
3	多功能演出厅	Multifunctional hall
4	乐器博物馆	Musical instruments museum
5	酒店	Hotel
6	排练厅	Rehearsal rooms

远眺黄海的鸟瞰图
Bird's-eye view looking toward the Yellow Sea

5

ates both logistical and functional synergies that offer the management and marketing of culture clear benefits. Like the principle of sports arenas, a grand theater combines not only sundry stages and corresponding genres of art, but can also be supplemented with other items such as a museum, shopping mall, or hotel. There are (almost) no limits to the dreams of clients or operators here, provided the "components" concerned are closely attuned to each other. The architecture then really has its job cut out for it, since it has to make an architecturally coherent whole from the wide functional variety. That is just what architects von Gerkan, Marg and Partners (gmp) have pulled off most impressively at the Qingdao Grand Theater.

Forging formal unity from a variety of expressive forms is an architectural leitmotif at gmp, featuring in their architectural thinking for many decades. Instead of wallowing in variety or indulging in an idiom of "anything goes" in terms of style and form, gmp deliberately deploys just a few signs and accents, but uses them all the more powerfully. In a musical analogy, this might be understood in terms of a "leitmotif", as used by German opera composer Wagner in his works, whereby a particular theme is picked up with subsequent repetitions and variations. In this sense, architecture would be "petrified music" (in Schelling's term) or "silent music" (Goethe). Its three-dimensional score is followed by the eye and the senses until structure and its location have been completely absorbed. Thus, visitors to the Qingdao Grand Theater discern certain principles–akin to a musical leitmotif–that are repeated, varied, reflect each other, or even cancel each other out. Architecture can be understood and even enjoyed as a composition in terms of its clearly developed structural themes.

Officially opened on 28 October 2010, Qingdao Grand Theater is couched topographically between the picturesque Lao Shan Mountains and the Yellow Sea in the eastern part of the city of Qingdao, which extends for some miles along the seashore. Thematically, the four separate structures–1. the opera House (1,600 seats), 2. the concert hall (1,200 seats) with its multifunction hall (400 seats), 3. musical instruments museum and 4. hotel–respond more to topographical features than urban ones. One could say, for example, that the Grand Theater stands above its urban context. It stands out from the shimmering office towers of the business district

and the mannered roof landscapes of the post-modernist residential blocks as a large white solid. Beyond all these constructional, stylistic flourishes, the Grand Theater intervenes symbolically between the forces of nature and symbolizes them–rocks and water. The rock surrounded (and thereby polished) by water is the thematic leitmotif of the architecture. The deliberate placing and positioning of the structure in the middle of a park stretching from the shore to the foot of the mountains picks up the theme and varies the process. The ground furrowed by water courses is the beginning of all landscapes, and in this way gains its characteristic appearance. The architects could have scarcely made the references to the place more relevant or distinctive. The Grand Theater is thus hallmarked with an immediately recognizable absence of ambivalence. The strength and power of the design resides in the recognizability associated therewith.

The variations of the leitmotif are asserted architecturally in a number of ways. Individual parts of the building, perspectives, glimpses of the interior, and outward views and the materiality all play an outstanding role.

6

7

相关的可识别性。

主乐调的变化通过各种方式在建筑设计上予以确定。各建筑单体组成部分、透视、室内一瞥、向外的视线和选材均起到了极为重要的作用。项目负责人尼古拉斯·博兰克与gmp柏林事务所的相关合作伙伴合作，负责整体设计和设计深化。"海浪和山体之间存在清晰的象征性联系，但同时我们也设法在空间上实现这一景观主题。外立面水平线条和屋顶百叶清晰地体现出线条的动感，并吸引人们走向空间深处。事实上，这种自由诠释的景观形态其实是建立在一种明确的几何结构之上的，即45cm的竖向建筑网格。这一网格在水平方向界定并连接了所有建筑组成部分，各结构构件仅在这些下沉线的位置转向消失或在其他地方重新实现连接。即使是台阶也设计为每级台阶15cm，这样每三级台阶的高度也正好符合这一结构。此外，在建筑的首层平面结构中，除了直角之外，只采用了一个15°的角度，且所有斜撑都符合这一体系。'沉降板'事实上控制了建筑的结构统一性。"

开放、弧线型的百叶屋顶结构将各单体建筑交织结合为一个整体。观众穿过宽阔的露天台阶，前往4.5m标高的平台，并由此前往大剧院其他功能分区；从象征意义上而言，人们或是从海滨或是从崂山的方向走向大剧院，然后穿越大剧院，所以他们的目光必然会越过平台，看见海滨或是山体这些自然景观的背景。从视觉上而言，人们将城市抛在身后，注目于大自然的壮丽景观，然后转向艺术和文化活动，这一设计手法从建筑设计上确立通往艺术之路。交通流线是场所感的组成部分，也有助于人们体验大剧院这一建筑，并为人们带来有意识的体验历程。线性的建筑结构呈锥形向顶端缩小，其表面覆盖采用山东出产的白麻花岗岩；这些线性结构构件与进深达6m，某些部位可长达100m，屋面元素所形成平行弧线也是建筑体验的重要组成部分。屋顶元素投射出阴影，而百叶的最终目的旨在营造出主要的空间效果。百叶将歌剧院/音乐厅与博物馆/酒店紧密联系起来，并从象征意义上将黄海和崂山联系起来。由于立面仅在几处设有玻璃幕墙以及众多重叠设置的细长窗户，线条的轻盈感觉可削弱立面的巨大体量和单一性。外立面的线条延伸宛如水平分层的沉积层，引导视线望向地

面柔和起伏的建筑。设计未设置任何尖角或转角，以避免破坏这一平衡结构所营造出的和谐氛围。尽管如此，建筑依然流露出应有的活力和优雅。尼古拉斯·博兰克解释说："由于建筑的几何形状已在掌控之中，因此，任何时候建筑的设计都可以遵循建筑内的一根线条，例如之前的平台层。正是这些水平的线条不断地将整个建筑群绑在一起。与gmp所设计的许多其他项目相比，青岛大剧院的基础从结构而言是非常牢固的，而且诠释又是相当自由的；根据这一基础，再加上几个精心设计的细节，就可以实现整个建筑的设计。这样的设计

8

思路使得设计方案具备可定义性和可施工性，并最终体现了青岛大剧院经典优雅的内涵。"

观众沿几段外部台阶拾级而上，即可进入歌剧院、音乐厅和多功能厅的中空大堂。多功能厅的装修纯粹从功能上考虑为一个黑盒子；而歌剧院和音乐厅的观众席则成为设计的亮点。歌剧院的观众厅沿用了水平设置的外立面结构的炫目白色，并与玻璃纤维钢筋混凝土内的红色/黑色竖向混凝土元素相互交替。这些厚重的、尺寸为10cm的混凝土元素，根据频带的要求而凿成不同的深度，确保实现良好的声学效果。观众厅的视觉效果充满欧洲歌剧院的节日氛围。音乐厅采用榆木覆层，显得更为轻松。上方的水平白色石膏板墙体覆层凸出，体现出

9

The responsible architects: "There is a clear symbolic correspondence between the waves of water and the mountain massif, but at the same time, we tried to implement this landscape theme spatially as well. The dynamic of line, readable in the horizontal façade lines and roof lamellae, draws you into the depth of the space. Admittedly, this freely interpreted landscape form is based on an unambiguous geometrical structure, i.e., a vertical building grid of forty-five-centimeters, which defines and connects all the buildings horizontally. The structures sheer off or connect up again elsewhere only at these sediment lines. Even the steps, each three of them fifteen centimeters high, fit into this structure. In addition, we have apart from right angles only a 15° angle in the ground plan structure of the building, and all diagonals are subject to this system. The 'sediment plates' literally control the structural unity of the building."

The buildings are interwoven into an ensemble by an open, curved linear lamella roof structure. Visitors are guided across broad open-air steps to a 4.5-meter-high terrace, from which all areas can be accessed. Figuratively, people come either from the sea or the mountains to and through the Grand Theater, and their eyes pass out of necessity across the terrace toward one or the other natural backdrop. Visually, you leave the city behind you, and turn to look at the grandeur of nature, and only then at art and culture. Access to art is thus specified by the architecture. The routes are part of the place, and help us to experience it. Visitors are thus subject to a conscious process. The linearity of the structures, which taper conically upwards and are clad with white Shandong Bei Ma granite, is as much a part of that as the parallel curve of the roof elements, which are up to six meters deep and in some cases over one hundred meters long. The latter cast shadows, but the ultimate purpose of the lamellae is the key spatial effect. They closely connect the opera house and concert hall with the museum and hotel, and perhaps also in figurative sense, the Yellow Sea with the Laoshan mountains. The lightness of the linearity offsets the mass and unity of the façades, which are interrupted by glass façades only in few places here and there, or numerous long, narrow window slits on top of each other. The lines of the façades run like horizontally layered sediments, and guide the eye along the softly polished building. No sharp angles or corners disrupt the harmony of this well-balanced structure. And even so, it exudes a due portion of dynamism and elegance, as the architects explains:
"Because of the controlled geometry, I can follow a line—which, for example, existed previously as a terrace level—in the building at any time. It is the horizontal lines that constantly pull the whole ensemble together. The structurally very strong foundation—which gmp, in comparison with many others of its buildings, interprets very freely—is the basis of the design. It only needs a few details to get the whole building. The result of that is a definability and constructability, but ultimately is what accounts for the classic elegance of the Qingdao Grand Theater."

After the external flights of steps, visitors enter the multi-story foyers of the opera house and concert hall and multifunctional hall. The latter is admittedly designed purely functionally as a black box. On the other hand, the auditoria of the opera house and concert hall are designed as showpieces. The dazzling white of the horizontally arranged façade structure is followed in the auditorium of the opera house by alternating vertical red and black concrete elements in glass-fiber reinforced concrete (GRC). The heavy, ten-centimeter elements are grooved to different depths according to frequency bands, and guarantee excellent acoustics. The visual impression is reminiscent of the festive atmosphere of European opera houses. The concert hall, with its elm veneer, is in contrast much lighter. The horizontal white wall-cladding above bows out three-dimensionally, and this is accentuated by appropriate lighting. Here too the fluid, undulating line is continued formally. Many of these sighting alignments run in both auditoria toward the stage.

10

三维的效果，应采用适当的照明加以强调；此处在形式上也继续沿用流动而起伏的线条。歌剧厅和音乐厅内流动的线条装饰都将人们视线引至舞台处。

然而在歌剧院和音乐厅中，声学效果比外观更为重要。在欧洲音乐剧院的历史中，有一些里程碑式的建筑，如瓦格纳在拜罗伊特设计的节庆大剧院、夏隆设计的柏林爱乐厅﹝图11﹞，均因与观众席设计直接相关的独特声学效果而备受推崇。建筑和音乐在歌剧院和音乐厅通过独特的形式和声学功能的组合而融合在一起。遗憾的是，在当今，由于舞台不再仅用于单一类型的艺术表演，这种建筑和音乐的完美融合已经极为罕见了。如今的文化产业发展要求灵活且差异性极大的技术和功能支持。因此，青岛大剧院的舞台也可同时用于各种截然不同的艺术形式。从纯技术的角度来看，青岛大剧院的舞台适合进行任何演出，从技术要求极高的欧洲歌剧到音乐剧、舞蹈演出和电视节目。对于歌剧院和音乐厅的观众席而言，其纯粹的声学尺寸，即非电声演出，是首要的设计考量。现代舞台须具备这一可变性，才能在吸引优秀艺术家和作品的竞争中不落人后。

在这一方面，青岛大剧院将能满足所有演员、制作人和观众的期待，并通过独有的建筑特色与多功能的实用性的协调统一，跻身中国最现代、最大气的大剧院之列，并使其自身成为一部能展示中西方传统歌剧、曲艺和音乐艺术的纯粹的建筑艺术佳作。

11

However, acoustics are much more important than looks in opera houses and concert halls. In the tradition of the European musical theater, there are certain milestones such as Wagner's Festspielhaus in Bayreuth or Scharoun's Berlin Philharmonic, which stand out for special acoustics directly associated with the design of the auditorium. Architecture and music fuse here in a unique combination of form and acoustic function—a state that is unfortunately ever rarer these days, as stages are no longer used for performances of a single artistic genre. Today's culture industry needs flexible and highly differentiated technical and functional backup. The stages of the Qingdao Grand Theater are therefore set up simultaneously for highly different forms of art. From a purely technical point of view, everything can be performed on the stage there, from the most technically demanding European opera to musicals, dance shows, and TV shows. In both auditoria, however, the purely acoustic dimension—i.e., unreinforced performance— was the prime design consideration. Modern stages have to provide this variety in order keep up with the competition for artistes and productions.

In this respect, the Qingdao Grand Theater will meet the expectations of all artists, producers, and audiences. In harmony with its remarkable architectural idiom and multifunctional practicality, it is one of the most modern and ambitious grand theaters in China—a work of total art where operas and music can be performed in accordance with both Chinese and European tradition.

← 演职人员以及乐器博物馆入口处。建筑体呈锥形向上收窄，流线型幕墙由白色花岗岩石材覆盖，细
长的窗带强调了其横向的肌理。
View of the artists' entrance and the Instrument Museum. The lines of the white
granite cladding, which conically reduce towards the top, are emphasized by the
narrow fenestration bands.

↑ 东南侧视角
View from the southeast

大剧院主入口。宽达6m，部分达100m长的屋顶构件营造了恢宏的空间。

← 大剧院主入口。宽达6m，部分达100m长的屋顶构件营造了恢宏的空间。
   View of the main entrance to the Opera House. The roof elements, which
   are up to 6 meter deep and over 100 meter long create a key spatial effect.
↑ 从北侧的室外阶梯处望向歌剧厅所在的建筑体块
   The northern open-air steps with view of the stage tower of the Opera House

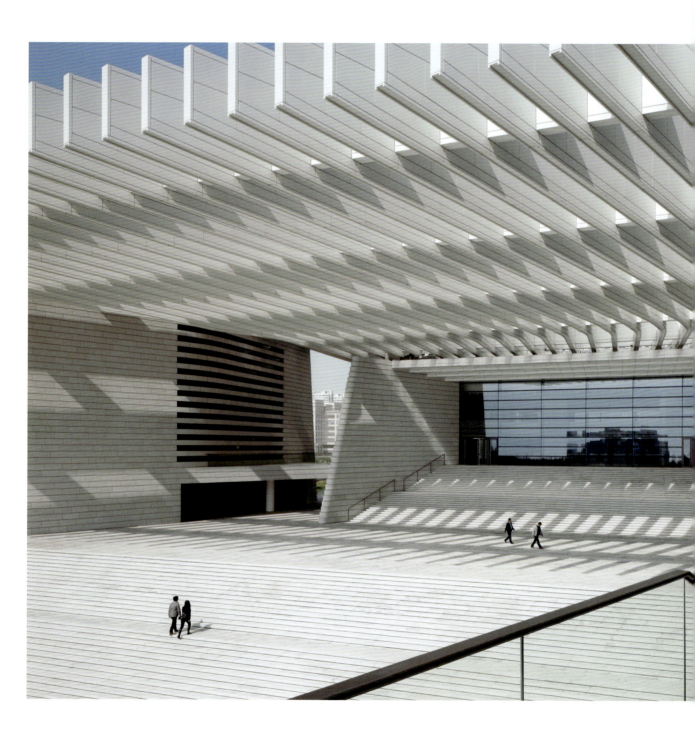

← ⊡ 音乐厅的休息平台拥有面向大海的视野
The concert hall's interval terrace with a view to the ocean

建筑的开放式弧线叶片状屋顶结构交织成为一个整体，引导观众跨过宽大的露天阶梯走上平台，并由那里进入各个功能区域。
The buildings are interwoven into an ensemble by an open, curved linear lamella roof structure. Visitors are guided across broad open-air steps to a terrace, from which all areas can be accessed.

↑   多功能演出厅的前厅以及通向平台处的楼梯

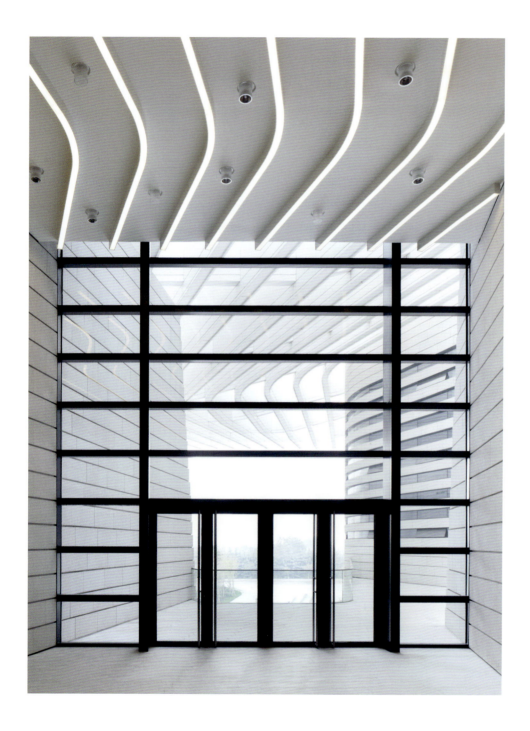

↑   叶片状流动的屋面与建筑体和谐统一，建筑内部和外部均为水平方向的立面肌理所刻画，令细长的
    窗带以及横向布局的玻璃幕墙得到了强调。
    The harmonious interplay of all building elements is emphasized by the sweeping
    parallel roof elements, the predominantly horizontal lines of the enclosing
    façades of the exterior and interior, the narrow recessed fenestration bands and
    the design of the glass façades.
→   多功能演出厅的前厅以及通向平台处的楼梯
    Foyer of the multifunctional hall with access to the balcony

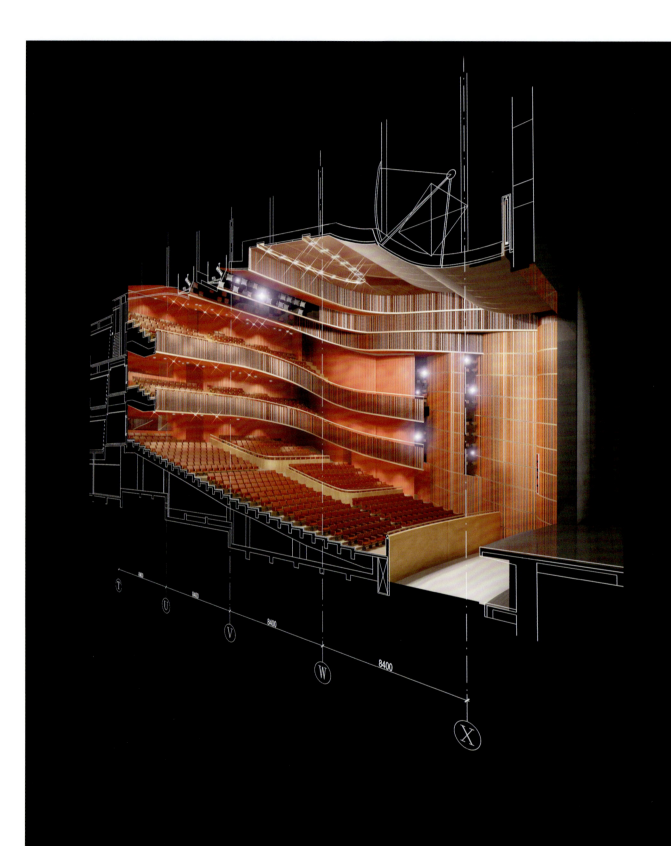

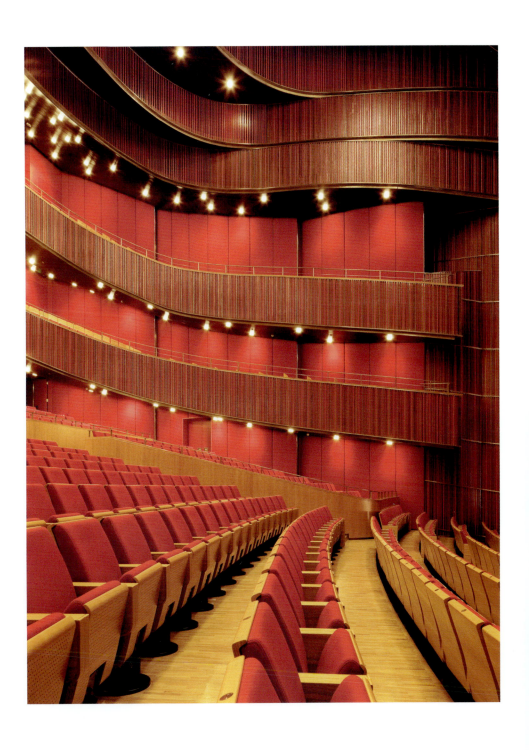

←　歌剧厅剖面透视
Sectional perspective of the Opera
House auditorium

↑　歌剧厅设有1 600个席位
The Opera House auditorium
has 1,600 seats.

←　▢▢ 歌剧厅前厅，与歌剧厅采用相同的混凝土饰面板装饰。
Foyer of the Opera House, which—like the auditorium—
is lined with sculptural concrete panels.

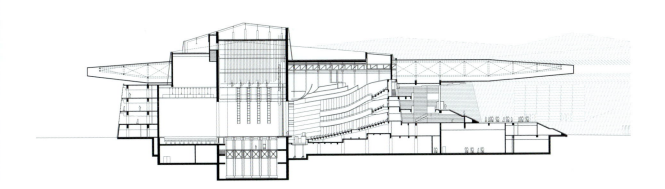

← 内壁采用不同材质的混凝土饰面板创造了满足国际标准的声学效果。
The sophisticated design of the sculptural concrete panels generates
acoustic conditions meeting international standards.

↑ 歌剧厅采取了欧洲歌剧院的经典格局, 位于多层之上的台座以马蹄铁的形式环绕舞台布置。
The auditorium is designed along classic European lines. The stalls are
surrounded by numerous balconies in the shape of a horseshoe.

↑ 歌剧厅横向剖面
Longitudinal section through the Opera House

↖ 歌剧厅内部再现了经典的欧洲歌剧院院面貌。
The visual impression is reminiscent of the festive
character of European opera houses.

↑ 与歌剧厅产生对比的音乐厅通过白色的吊顶和榆木镶板墙壁呈现出更加轻
松愉悦的空间氛围。
In contrast to the Opera House auditorium, the white
ceiling and elm wood veneer of the concert hall stalls
impart a cheerful character.

↑　多边形音乐厅内设有
1 200个坐席。
The layout of the audito-
rium is diamond-shaped
and has seating for 1,200.

↗　白色的石膏墙壁饰面板呈立体波浪形错落布局，实现完美的声学品质。
The white plasterboard lining of the walls recedes and pro-
trudes to produce a three-dimensional effect and enhances
the high quality of the concert hall acoustics.

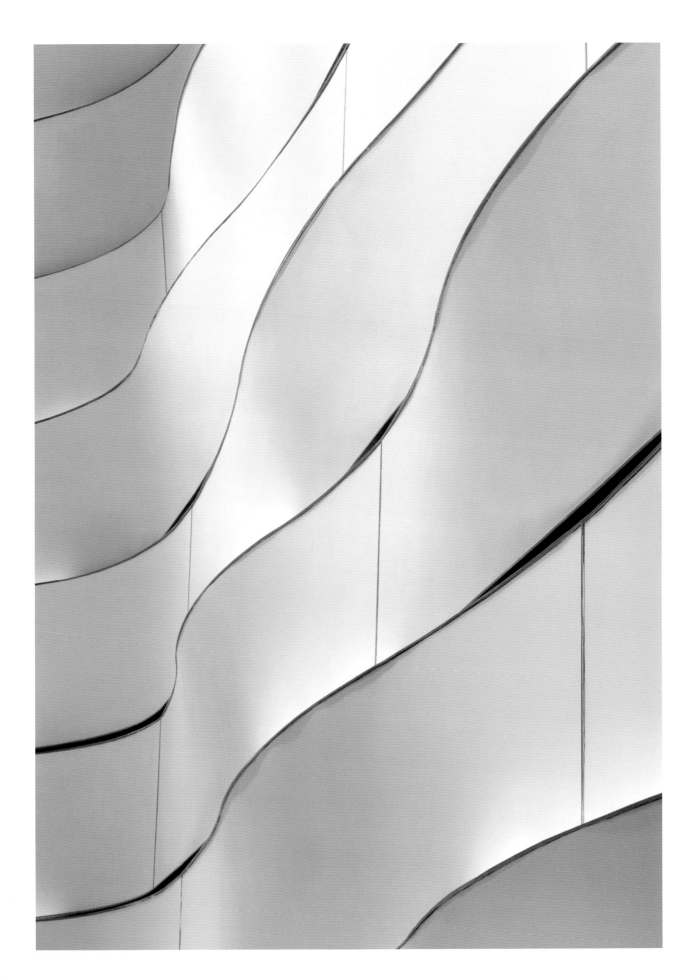

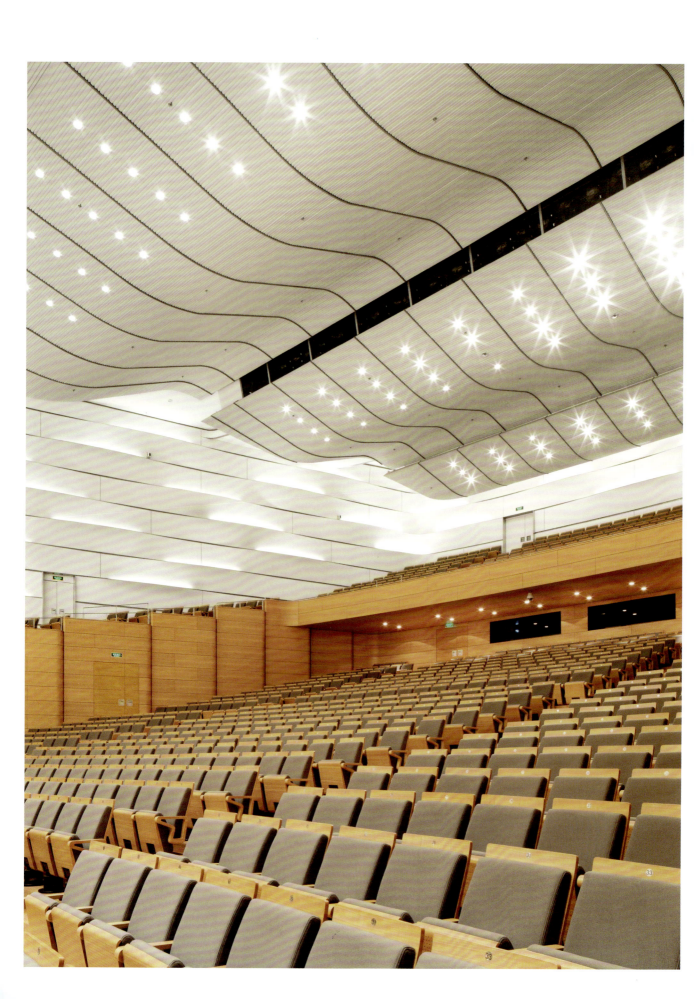

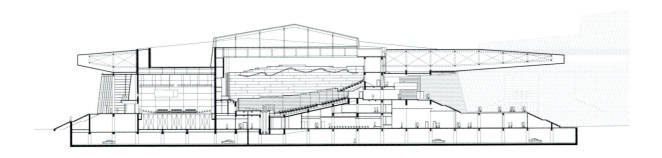

← 照明刻画建筑形象。
The lighting has been arranged to accentuate the building components.

↑ 波浪状的造型除符合声学要求外还传达了灵动起伏的音乐海洋的意象。
The sculptural, wavy design of the building elements contributes to the excellent acoustics, while also symbolizing the ocean surrounding Qingdao.

↑ 音乐厅横向剖面
Longitudinal section of the concert hall

↑   内壁造型保证了理想的声学效果。
All building components have a
sculptured design in order to generate
a balanced acoustic effect.

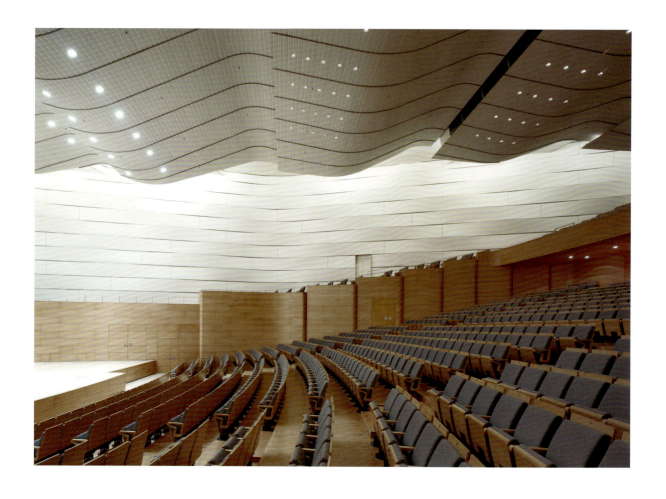

↑   内壁造型保证了理想的声学效果。
All building components have a
sculptured design in order to generate
a balanced acoustic effect.

→   音乐厅剖面透视
Sectional perspective
of the concert hall

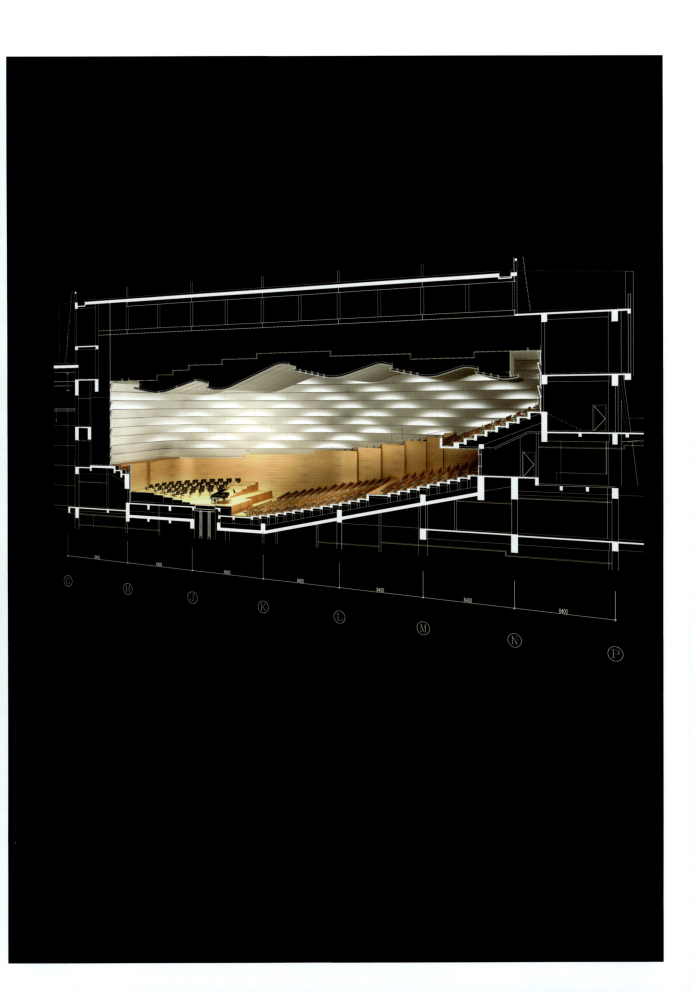

↑   可拆卸坐席可容纳400名观众
There is mobile seating for up
to 400 visitors.

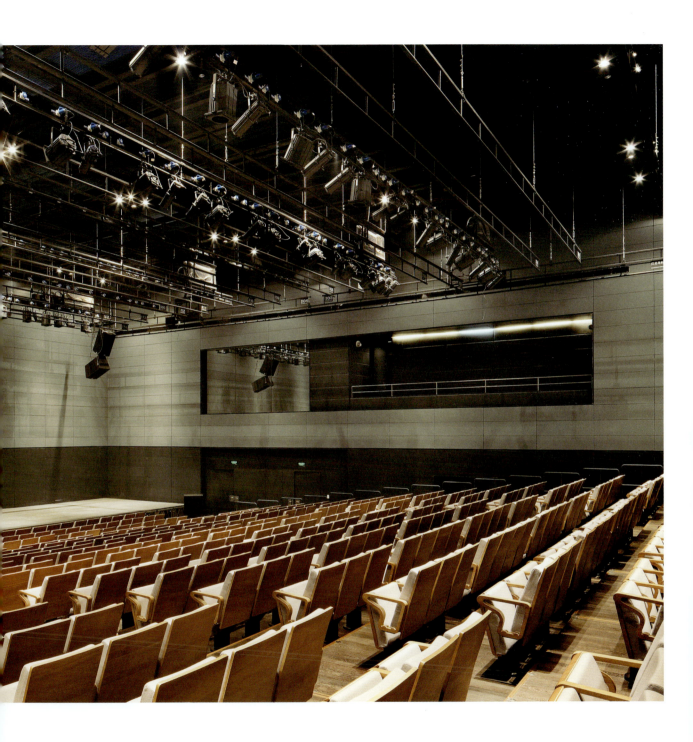

↑ 被塑造成"黑盒子"的多功能演出厅可满足多种用途。
The multifunctional hall has been designed as a "black box"
and can accommodate a wide range of functions.

在多层挑空的前厅内，来访者可以感受到流动屋面元素以及
白色花岗岩刻画出的横向肌理在室内设计中的延续。
The parallel curves of the roof elements
and the horizontal structure of the light granite
can also be experienced by visitors in the
multistory foyer.

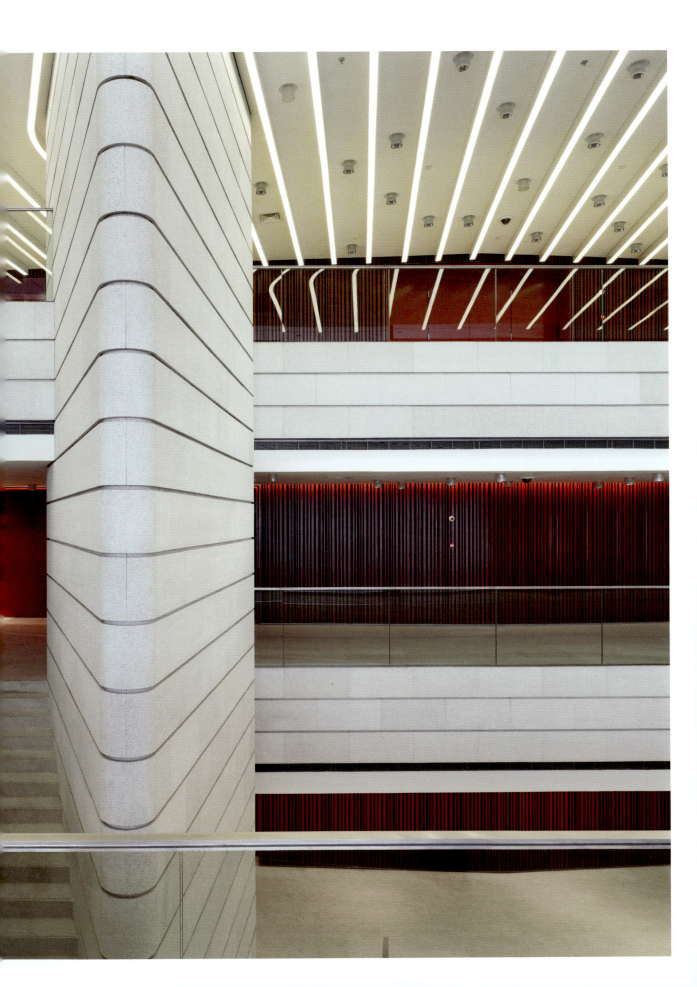

从通廊处观望歌剧院前厅，通廊连接了南北侧广场。
View into the foyer of the Opera House along the passage linking the northern and southern concourses.

从通廊处观望歌剧院前厅，通廊连接了南北侧广场。
View into the foyer of the Opera House along the passage linking the northern and southern concourses.

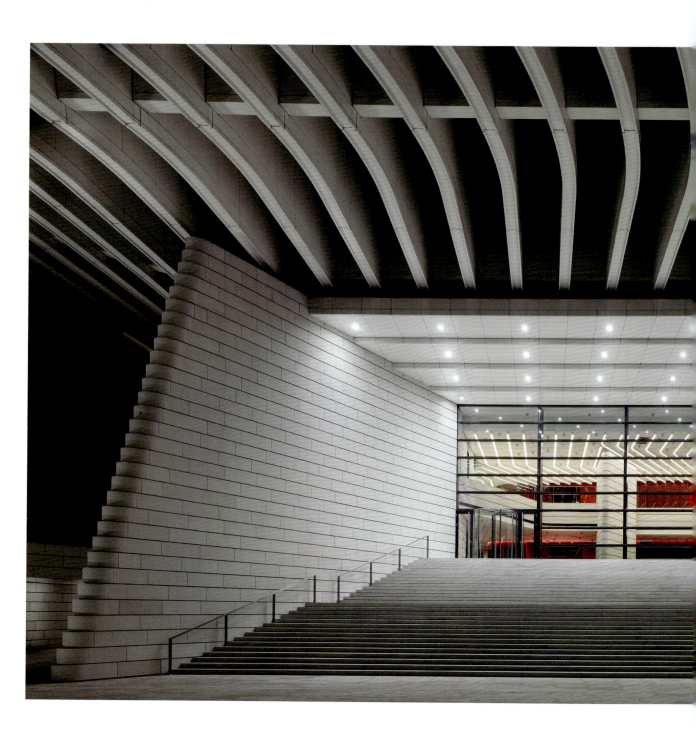

大剧院主入口照明效果
Main entrance to the Opera House
with festive lighting

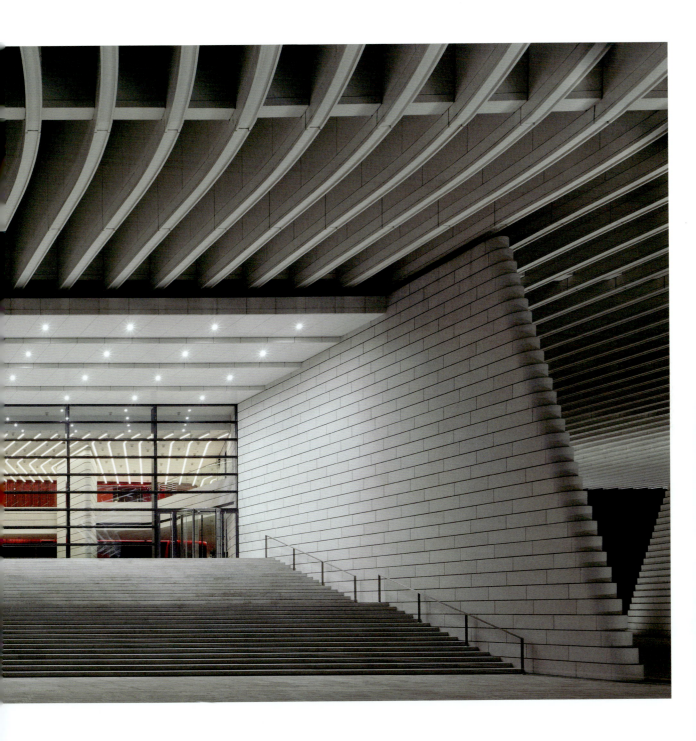

南侧前广场上的大剧院照明效果

南侧前广场上的大剧院照明效果
View of the illuminated southern concourse

说明 Imprint

总策划 Concept
迈克尔·库恩
德国冯·格康，玛格及合伙人建筑师事务所（gmp）
公共关系和新闻出版部负责人
Michael Kuhn (gmp)
Head of PR
柳青
《城市·环境·设计》（UED）杂志社执行主编
Liu Qing (UED)
Executive Chief Editor of UEDmagazine

编辑 Editing Direction
克劳迪娅·苔斯勒 Claudia Tiesler (gmp)
方小诗 Fang Xiaoshi (gmp)
郑珊珊 Zheng Shan Shan (Peking gmp)

平面设计 Layout and Typesetting
汤姆·魏伯伦茨、亨德里克西什莱
欧恩平面设计公司（德国汉堡）
ON Grafik – Tom Wibberenz mit with
Hendrik Sichler, Hamburg

校审 Proofreading
克劳迪娅·苔斯勒 Claudia Tiesler (gmp)
方小诗 Fang Xiaoshi (gmp)
包伸明 Bao Shenming (UED)
姜思琪 Jiang Siqi (UED)

翻译 Translation
哈特温·布什（英国阿什顿）
Hartwin Busch, Ashdon, UK (en.)
克里斯蒂安·布莱辛（德国柏林）
Christian Brensing, Berlin (en.)
方小诗 Fang Xiaoshi (chin.) (gmp)

图片处理 Picture Editing
特里克茜·汉森 Trixi Hansen (gmp)
吉多·布里克斯纳 Guido Brixner (gmp)

印刷制作 Print Production and Binding
北京雅昌彩色印刷有限公司
ARTRON